The Little Fun Book of Molecules/Humans

Molecules/Humans

By

John Hodgson

ISBN: 1-4033-8524-6 (e-book)
ISBN: 1-4033-8525-4 (Paperback)

Library of Congress Control Number: 2002095155

This book is printed on acid free paper.

Printed in the United States of America
Bloomington, IN

1stBooks - rev. 10/30/02

PREFACE

This book is intended to look at the similarities existing between two entities. If we understand more of each entity there may be clues to hold new clues to hold new scientific information leading to new research.

Molecules/humans are prone to disease. Molecules/humans cannot fight off some diseases without outside help.

John Hodgson

Molecules/humans don't always get along. That's why they don't come together.

Different molecules/humans behave differently having different reaction/behavior to changing chemicals/situations.

John Hodgson

In the world molecules/humans are different/equal compared to each other.

In communities, molecules/humans differ/gather with each other joining/making friends.

John Hodgson

When molecules/humans play they move quickly/play basketball doing what is needed to form cells/shoot a basket.

Molecules/humans form substances/make things to form life/develop.

John Hodgson

When   molecules/humans
need/stretch the imagination they
form/develop.

When molecules/humans mesh they have chemical or physical reaction/produce.

John Hodgson

Molecules/Humans share from one another taking what they need from each other then separating.

Good molecules/humans excel and teach one another. Bad molecules/humans don't interact and fall down.

John Hodgson

Molecules/humans are different sizes and shapes. Molecules/humans attract one another.

There are many molecules/humans on earth. Each with it's own distinction. Some molecules/humans are bad/good.

John Hodgson

Some molecules/humans produce offspring. Some more than others, some doubles and triplets.

When molecules/humans breathe they accept oxygen. Turning the oxygen to useful energy, then living long and prosperous.

Molecules/humans need each other to form new substances/get married.

Molecules/humans group together/play hockey to make solvents/have fun.

John Hodgson

When molecules/humans come in contact some get along/bond and form substances/relationships to last a lifetime.

With experiment we can understand these molecules/humans like we never knew before.

John Hodgson

Why do molecules/humans behave the way they do?

Molecules/humans sometimes devour/ step on each other to get what they want.

Molecules/humans   get
hyperactive when fed certain
proteins/sugar.

Molecules/humans pair together to create different chemicals/babies and fall in love.

John Hodgson

Molecules/humans have cancer/war destroying many molecules/ humans.

Some Molecules/humans become cells/friends while other molecules/humans do not like each other/mixtures.

John Hodgson

Some molecules/humans transfix on each other when in contact/speak.

We must find out about Molecules/humans so we can understand disease/cures.

The molecules/humans/cells live on earth forming new complex molecules/offspring.

Some molecules/humans are more successful in the world/have more blessings.

John Hodgson

Molecules/humans are combined with each other forming animals/nations.

Complex molecules/humans know plenty of the same elements/creatures.

There are molecules/humans that go to experiments/school to under stand each molecule/human.

Each element/human needs a nucleus/teacher to live, develop and learn.

An element/pupil gives a nucleus/teacher information that is needed for development/body.

Without the Molecules/humans living/learning there would be no life/future.

Us Molecules/humans must understand the similarities between each other.

Certain Molecules/Humans defend each other like disease/war.

Many molecules/humans have good relationships/fulfill each others wants and desires.

Molecules/humans that pair together have love/positive relationships.

John Hodgson

Molecules/humans that enjoy each others company form solutions/enjoy things together.

Molecules/humans have needs to bond/love.

John Hodgson

Molecules/humans needs are amazing/tremendous.

To understand molecules/humans we must do experiments/learn from each other.

John Hodgson

Some complex molecules/humans make special relationships/get married.

Some molecules/humans never find a special relationship/remain on their own.

John Hodgson

Some molecules/humans never get to bond/get what they want.

Some molecules/humans show their true colors/get what they need.

Molecules/humans need each other/get lonely.

Molecules/humans   lay
dormant/sit down.

John Hodgson

Molecules/humans can get along/do not always see together eye to eye.

When molecules/humans do not think alike/might separate.

Some molecules/humans like each other and bond/leave each other bitter.

Some molecules/humans travel with others/travel alone.

John Hodgson

Some molecules/humans join together/liking each others company.

Molecules/humans join each other forming solutions/communicate.

Molecules/humans like to bond with things they like/do different things.

Molecules/humans push each other around/bully one another.

Some molecules/humans get what they want/ go after what they want.

Molecules/humans take in elements/food.

John Hodgson

Molecules/humans engage in different behavior/sex.

Molecules/humans develop to be strong/men.

John Hodgson

Molecules/humans forge new innovation/have specific goals.

Molecules/humans enrich one another/changing with time.

John Hodgson

Molecules/humans employ each other/have standards.

Molecules/humans have special needs/food water.

John Hodgson

Molecules/humans   make
change/common bonds.

If molecules/humans try hard to get what they need, molecules/humans will find that there is beauty/peace.

If molecules/humans can understand what they need from each other and not what they want from each other there would be a symbiotic relationship/nurturing relationship.

Molecules/humans move around to different molecules/humans/migrate.

John Hodgson

Living molecules/humans need nutrients/food.

Molecules/humans   form
chains/dance.

John Hodgson

Molecules/humans are complicated/sophisticated.

Molecules/humans   need energy/work.

John Hodgson

Without   nutrients,
molecules/humans   will
die/suffer.

Molecules/humans do travel/move/migrate to one another.

John Hodgson

Molecules/humans feed off one another   having relationships/parenting.

Molecules/humans live from one another throughout their lifetime.

John Hodgson

Each Molecules/human is like nucleus/earth living with each other. So humans must understand the earth we all live on.

Molecules/humans like the goodness that is felt from each other/get married.

John Hodgson

Molecules/humans   have
memories/sometimes   ignore
each other.

Molecules/humans know what they need or get what they need/find peace to be wonderful.

John Hodgson

Molecules/humans   forming
complexities/going to concert.

Molecules/humans agree with each other forming bonds/friends.

John Hodgson

Certain molecules/humans attract one another/elements/marriage.

Molecules/humans act much alike given the right elements.

John Hodgson

Molecules/humans fall into many categories/occupations.

Molecules/humans exist to better life/live longer.

John Hodgson

Molecules/humans are living to find similarities/experiments.

Molecules/humans hold answers to benefit all/learn.

John Hodgson

Molecules/humans   shouldn't
give up/hope.

Molecules/humans have power/with speech/words.

John Hodgson

Molecules/humans   should
participate/live.

Molecules/humans are forever evolving/finding new theories.

John Hodgson

Molecules/humans   are
changing/evolve.

Molecules/humans   govern
life/harness life.

John Hodgson

Molecules/humans are in development/evolving.

Molecules/humans   get
frustrated/finding cures.

John Hodgson

Molecules/humans hold answers to questions/experiments.

THE END

John Hodgson

About the Author

He enjoys spending time learning new concepts using knowledge he has learned through studying relationships and interactions of human behavior in society and workplace.